HIGHWAY CONSTRUCTION AND INSPECTION FIELDBOOK

HIGHWAY CONSTRUCTION AND INSPECTION FIELDBOOK

Project Construction Management Book

ALBERTO MUNGUIA MIRELES

HIGHWAY CONSTRUCTION AND INSPECTION FIELDBOOK
PROJECT CONSTRUCTION MANAGEMENT BOOK

iUniverse books may be ordered through booksellers or by contacting:

iUniverse
1663 Liberty Drive
Bloomington, IN 47403
www.iuniverse.com
1-800-Authors (1-800-288-4677)

ISBN: 978-1-4917-4739-1 (sc)
ISBN: 978-1-4917-4741-4 (hc)
ISBN: 978-1-4917-4740-7 (e)

Library of Congress Control Number: 2014917288

Printed in the United States of America.

iUniverse rev. date: 10/29/2014

Highway Construction and Inspection Fieldbook

Project Name: _____

_____

City: _____

County: _____

Contractor: _____

Inspection Company: _____

Contact Information

Name: _____

Phone no.: _____

Book No. ___ **of** ___
Month-Year:_____

Receive free training on subjects related to highway
construction at www.cs4highway.com.

Contents

A Note to the Reader

This fieldbook will help you to document daily activities and critical events properly on the job site. Its guidelines and key factors for writing effective daily reports are based on real-world results. The procedures were created by analyzing the most effective methods used by succesful contractors and over fifty thousand hours of hard-won experience dealing with project management during initiating, planning, executing, monitoring, controlling, and closing groups during my twenty-four-year career of providing civil engineering services during the contruction and inspection of infrastructure projects.

This fieldbook is a tested methodology that, if used well and daily, will provide consistent information to the user as well as to upper management. The information collected in this fieldbook will aid supervisors and upper management in tracking costs, monitoring and improving the uses of resources, and documenting how a project is developing in compliance with contractual documents.

This is not like a common project management book—intended to be read and then stored on a shelf. *It contains blank sheets to enter information on daily operations for a full month,* so it is intended to be carried out to the job site daily. But it is also a book. *It is a source of information and training about how to writte effective daily reports.* It is, then, a *fieldbook* that includes procedures used by roadwork experts and improved by my abilities, competences, and experience, gained through years of doing this work.

You might wonder what led me to begin the endeavor of writing a book and providing related training on my website. Well, it is not a secret that the Texas Department of Transportation (TxDOT) has been adopting a different role in the construction of its state highway system. Nowadays, local governments are funding their projects through different types of programs that allow them to construct them earlier than they would through traditional programs.

Many private companies have emerged to take advantage of these new opportunities, but they are facing a tremendous shortage of capable and well-trained inspectors and project managers that possess the knowledge and experience to perform new projects at the level of detail and quality that TxDOT specifications demand. In my opinion, one of the reason for this is simple: TxDOT used to have training programs that allowed them to teach their personnel and to be able to supply inspectors—though often even that was not enough. This created a need for innovative thinking to transfer the experience of qualified people to less-experienced personnel.

This is where I fit in. I have created a series of fieldbooks, Excel programs, PDF documents, templates, videos, and web seminars with the intention of creating good content that will help less-experienced personnel to obtain some of the training they need to succeed. Company owners will benefit greatly from these pages, because employees can obtain real-world situations training that I am sure will help them to be more effective in delivering good-quality projects.

Please visit my web page at www.cs4highway.com. Register to receive free training on subjects related to highway construction as well as to keep updated as more information becomes available. This website is a work in progress, so please bear with me if you see new pages under construction

It is my mission to instruct young engineers, project managers, students, contractors, subcontractors, supervisors, and inspectors who are dedicated to the road construction industry on how to utilize available tools and techniques to provide consistency in professional inspection services and to document construction processes properly to ensure that each project provides the service for which it was undertaken.

It is my belief that, having adopted high standards of behavior and dedication to personal fulfillment and to helping others attain it, I am in a position to help advance our industry to continue the delivery of projects in conformance with specifications, balancing the constraints of quality, scope, budget, schedule, and stakeholders.

How to Use This Fieldbook

This fieldbook is designed for entering information on daily operations during one full month. The daily report for a given date is made up of four templates:

1. Tracking equipment on-site
2. Manpower and summary of work performed
3. Measurements and calculations
4. Design-construction narrative

The framework described in this fieldbook supports the four templates, which *you need to fill out* to ensure that every day during the life cycle of the project you collect essential information in the systematic way required to satisfy federal and states audits.

Please understand that this is only one of the important documents you are required to maintain, but many other factors also contribute to the project's success. I am giving you the formula that will help you succeed on your quest of writing effective daily reports. But you are going to have to demand more of yourself.

For now, understand that daily reports as well as other types of communications written by construction managers, inspectors, and contractors help us to ensure the timely and appropriate generation, collection, dissemination, storage, and deposition of project information.

This book is divided into seven lessons and two sections. In each lesson, you will learn important information on writing effective daily reports. The sections include blank templates you will fill out to create your daily reports for the month.

You will also find two real-world highway construction examples to give you a glimpse of the use of the fieldbook templates. You may access more examples, additional training videos, and resources at www.cs4highway.com when you subscribe to that site.

Example 1

How to Fill Out the Four Templates in a Daily Report

It's a normal ninety-five-degree summer day in Texas, and the contractor is performing excavation and embankment operations at different locations along seventeen miles of the road. Controlling earthmoving operations is not a simple activity. Many processes have to be monitored and controlled to keep the operations under control. The control forms listed below are used by roadwork experts, as is the template that I am about to explain. (Visit www.cs4higway.com to find more information and to download the forms.)

- Number of loads counted by the contractor in carrying vehicles
- Possible production predicted by inspection VS load counts by earthmoving contractor
- Worked days on cuts and fills
- 110-2001 roadway excavation or 132-2006 roadway embankment breakdown forms
- Work report for excavation or embankment
- Schedule performance index

Monitoring Earthwork Operations—An Inspector's Perspective

I am the dirt inspector on-site, and it is time for me to drive to the job and inspect it. Previously, from the contract documents, I have identified that these activities will meet the requirements stated on items 132-2006, 110-2001, and 160-2003. I have also read the requirements in the specification, general notes, and special provisions. I clearly understand what is required to be done. I take my sets of drawings and jump into my four-by-four truck. It is a wonderful day to give the best in me.

As follows, you will find my daily report on the four templates:

1. Tracking equipment on-site
2. Manpower and summary of work performed
3. Measurements and calculations
4. Design-construction narrative

TEMPLATE 1.
TRACKING EQUIPMENT ON-SITE

EQ	U	I	B	CUT / Pumping FROM	TO	FILL FROM	TO	Model or Notes
Compactors		•						Cat 815
	1					332	334	Cat 825
	1					170	175	Cat 825
Compresors								
Water Pump	1			210	210+50			6" pump
Cranes								
Excavators	1			Coppland Pit				Cat 345 B
			X					Cat 345 C
Graders	1			Priefert Pit		320	334	Cat 14M
	1					258	264	Cat 14M (Top soil)
Loader		•						Cat 950 G
Scraper	1			Priefert	Pit	332	334	Cat 631E
	1			133	179	170	171	Cat 631E
		••						Cat 631E
	11			178	180	172	175	Cat 631G
	11			Priefert	Pit	330	332	Cat 631G
	11			Priefert	Pit	324	330	Cat 631G
	1111	•••		Coppland	Pit	306	308	Cat 740
Trucks	1			Coppland	Pit	306	308	J.Deer 400
	1				service			Fuel truck
	1					306	308	Cat 740 with Tank
Tractors	1			Priefert	Pit			D10R (Pushing Scr)
	1					167	168	D6
	1			177	180			D10N Pushing Scr.
	1					305	307	D8R
						170	171	D8R
		•						D8R
Scraper	1			352	354	332	334	Cat 623G
	1			308	310	306	307	Cat 623G
	1			260	263	R.o.w		Cat 623 Top Soil
Mixer								
Disc	1					170	173	Case STX 375
		•						Case stx 375
Roller Blade		•						Holmes 16D

TEMPLATE 2.
MANPOWER AND SUMMARY OF WORK PERFORMED

Date: 7-5-12 (Day) Night | S | M | T | W | (T) | F | S

Qty	Job Position	Qty	Job Position	Qty	Job Position
	Backhoe Operator		Front End Loader Operator	~~||||~~ 1	Scraper Operator
	Broom or Sweeper Op.	1	Hydraulic Excavator Op.	1	Serviceman
~~||||~~	Bulldozer Operator		Laborer		Steel Setter
	Concrete Finisher	~~||~~ 1	Leadman	11	Surveyor
	Concrete Saw Operator		Mechanic	1	Tractor Operator
	Carpenter		Mixer Operator	~~||||~~ 11	Truck Driver
	Crane Operator	11	Motor Grader Operator		Utility Labor
	Flagger		Pipelayer		Work Zone Barricades
1	Foreman	11	Roller Operator		
	Form Builder/Setter Struc				

Description of Work Performed

Activity	1	2	3	4	5	
Job No.	475	475	475			
Category	132	110	160			Total
Function	2006	2001	2003			
Name	Number of hours worked on each activity					
Thomas Gaytan (Crew 1)	10		2			12
Roger Cumming (Crew 2)	5	5	2			12
Jose Varela (Crew 3)	6	6				12

Activity	Description
1	Embankment (Final) Type c
2	Excavation (Roadway)
3	Furnishing and placing Topsoil 4"
4	
5	

TEMPLATE 3.
MEASUREMENTS AND CALCULATIONS

MEASUREMENTS AND CALCULATIONS							
DESCRIPTION	LOCATION	QTY	LENGTH (Ft)	WIDTH (Ft)	HEIGHT (Ft)	CONV FACTOR	VOLUME OR AREA
Over excavation. 303+10 to 303+90		1	80	20	2	1/27	118.51cy

PRODUCTIVITY RATE ANALISYS FOR PAYMENT PURPOSE

Crew no 1: (4e) Dump truck 740D + (1ea) EXC. 345

Measured time to cut = 1 minute, 45 sec.
Measured time to haul = 5 min - 45 sec.

Expected time to cut = 1.76 min
Expected time to haul = 5.79 min

Number of haul units required = $\frac{5.79}{1.76}$ = 4 trucks

  verification:
Trucks required = Trucks used → Good ✓

Cycle per hour = $\frac{60}{5.79}$ = 10 cycle per hour

Dump truck production rate = 10 cycles × $\frac{12 cy}{truck}$ = 120 cy/hr per truck.

Efficency factor = 85%
→ 120 × 0.85 = 102 cy/hr.truck

Number of hours worked In a day = 10 hrs

$\frac{102 cy}{hr.truck}$ × 10 hr = 1020 cy/truck

$\frac{1020 cy}{day.truck}$ × 4 trucks = 4080 cy/day

Volume to compare with load count provided by contractor

FOCUS POINTS

1	Weather statement : Fair , Max=95F/Min 66	⑦	Unusual construction work conditions	
②	Results of surveillance : Satisfactory	8	Rework or field errors	No
③	Important verbal instructions received or given	9	Warranty work done	No
4	Names of official visitors and summary of discussion	10	Customer adjustments	No
5	Result of safety inspection : Satisfactory	11	Estimating errors	No
⑥	Testing performed Densities	12	Pay work item	Yes

Date: 7-5-2012

Days remaining : 165
Number of the day of the year 186

2	Contractor continued cut and Fill operations in an orderly and systematic manner. Establishing cycles. Excavated material has been hauled and dumped over the embankment that was simultaneosly re-compacted. The dumped material is moved by dozing operations to incorporate it in uniform layers. Abutting layers of dissimilar materials have been mixed. the embankment is being done in layers parallel to the finish grade for the full width of the roadway section. the layer thickness during placement was inspected. contractor was adviced to keep it under requirement. GPS operator was seen on site placing construction stakes color coded. the material excavated meets the req. for type c embankment. It was observed than the excavation and embankment operations are being executed in a manner that permanent lines and grades of ditches are defined.
3	Jose V. showed me an area that migth be required to be over excavated e. sta 258+50. He will provided me a cross section.
6	Density taken e sta 324+50 was sattisfactory see density log.
7	Water is coming out from back slope @ sta 258+60 Rock rip rap may be required.
3	I contacted to our Project Manager fro inform him about the problem. I requested directions to proceed.

Section 1

Filling Out Daily Work Reports

EQ	QTY			CUT/		FILL/		Model or
	U	I	B	FROM	TO	FROM	TO	Notes
Compactors								
Compresors								
Cranes								
Excavators								
Graders								
Loader								
Man lift								
Trucks								
Tractors								
Scraper								
Mixer								

ALBERTO MUNGUIA MIRELES

Date:		Day	Night	S	M	T	W	T	F	S

Qty	Job Position	Qty	Job Position	Qty	Job Position
	Backhoe Operator		Front End Loader Operator		Scraper Operator
	Broom or Sweeper Op.		Hydraulic Excavator Op.		Serviceman
	Bulldozer Operator		Laborer		Steel Setter
	Concrete Finisher		Leadman		Surveyor
	Concrete Saw Operator		Mechanic		Tractor Operator
	Carpenter		Mixer Operator		Truck Driver
	Crane Operator		Motor Grader Operator		Utility Labor
	Flagger		Pipelayer		Work Zone Barricades
	Foreman		Roller Operator		
	Form Builder/Setter Struc				

Description of Work Performed						
Activity	1	2	3	4	5	
Job No.						Total
Category						
Function						
Name	Number of hours worked on each activity					

Activity	Description
1	
2	
3	
4	
5	

HIGHWAY CONSTRUCTION AND INSPECTION FIELDBOOK

MEASUREMENTS AND CALCULATIONS							
DESCRIPTION	LOCATION	QTY	LENGTH	WIDTH	HEIGHT	CONV FACTOR	VOLUME OR AREA

ALBERTO MUNGUIA MIRELES

FOCAL POINTS

1	Weather statement	7	Unusual construction work conditions
2	Results of surveillance	8	Rework or field errors
3	Important verbal instructions received or given	9	Warranty work done
4	Names of official visitors and summary of discussion	10	Customer adjustments
5	Result of safety inspection	11	Estimating errors
6	Testing performed	12	Pay work item

Date: Construction's days remaining

EQ	QTY			CUT/		FILL/		Model or
	U	I	B	FROM	TO	FROM	TO	Notes
Compactors								
Compresors								
Cranes								
Excavators								
Graders								
Loader								
Man lift								
Trucks								
Tractors								
Scraper								
Mixer								

Date:		Day	Night	S	M	T	W	T	F	S

Qty	Job Position	Qty	Job Position	Qty	Job Position
	Backhoe Operator		Front End Loader Operator		Scraper Operator
	Broom or Sweeper Op.		Hydraulic Excavator Op.		Serviceman
	Bulldozer Operator		Laborer		Steel Setter
	Concrete Finisher		Leadman		Surveyor
	Concrete Saw Operator		Mechanic		Tractor Operator
	Carpenter		Mixer Operator		Truck Driver
	Crane Operator		Motor Grader Operator		Utility Labor
	Flagger		Pipelayer		Work Zone Barricades
	Foreman		Roller Operator		
	Form Builder/Setter Struc				

Description of Work Performed						
Activity	1	2	3	4	5	
Job No.						Total
Category						
Function						
Name	Number of hours worked on each activity					

Activity	Description
1	
2	
3	
4	
5	

MEASUREMENTS AND CALCULATIONS							
DESCRIPTION	LOCATION	QTY	LENGTH	WIDTH	HEIGHT	CONV FACTOR	VOLUME OR AREA

ALBERTO MUNGUIA MIRELES

FOCAL POINTS

1	Weather statement	7	Unusual construction work conditions
2	Results of surveillance	8	Rework or field errors
3	Important verbal instructions received or given	9	Warranty work done
4	Names of official visitors and summary of discussion	10	Customer adjustments
5	Result of safety inspection	11	Estimating errors
6	Testing performed	12	Pay work item

Date: Construction's days remaining

EQ	QTY			CUT/		FILL/		Model or
	U	I	B	FROM	TO	FROM	TO	Notes
Compactors								
Compresors								
Cranes								
Excavators								
Graders								
Loader								
Man lift								
Trucks								
Tractors								
Scraper								
Mixer								

ALBERTO MUNGUIA MIRELES

Date:		Day	Night	S	M	T	W	T	F	S

Qty	Job Position	Qty	Job Position	Qty	Job Position
	Backhoe Operator		Front End Loader Operator		Scraper Operator
	Broom or Sweeper Op.		Hydraulic Excavator Op.		Serviceman
	Bulldozer Operator		Laborer		Steel Setter
	Concrete Finisher		Leadman		Surveyor
	Concrete Saw Operator		Mechanic		Tractor Operator
	Carpenter		Mixer Operator		Truck Driver
	Crane Operator		Motor Grader Operator		Utility Labor
	Flagger		Pipelayer		Work Zone Barricades
	Foreman		Roller Operator		
	Form Builder/Setter Struc				

Description of Work Performed						
Activity	1	2	3	4	5	
Job No.						Total
Category						
Function						
Name	Number of hours worked on each activity					

Activity	Description
1	
2	
3	
4	
5	

MEASUREMENTS AND CALCULATIONS							
DESCRIPTION	LOCATION	QTY	LENGTH	WIDTH	HEIGHT	CONV FACTOR	VOLUME OR AREA

ALBERTO MUNGUIA MIRELES

FOCAL POINTS

1	Weather statement	7	Unusual construction work conditions
2	Results of surveillance	8	Rework or field errors
3	Important verbal instructions received or given	9	Warranty work done
4	Names of official visitors and summary of discussion	10	Customer adjustments
5	Result of safety inspection	11	Estimating errors
6	Testing performed	12	Pay work item

Date: Construction's days remaining

EQ	QTY			CUT/		FILL/		Model or
	U	I	B	FROM	TO	FROM	TO	Notes
Compactors								
Compresors								
Cranes								
Excavators								
Graders								
Loader								
Man lift								
Trucks								
Tractors								
Scraper								
Mixer								

ALBERTO MUNGUIA MIRELES

Date:		Day	Night	S	M	T	W	T	F	S

Qty	Job Position	Qty	Job Position	Qty	Job Position
	Backhoe Operator		Front End Loader Operator		Scraper Operator
	Broom or Sweeper Op.		Hydraulic Excavator Op.		Serviceman
	Bulldozer Operator		Laborer		Steel Setter
	Concrete Finisher		Leadman		Surveyor
	Concrete Saw Operator		Mechanic		Tractor Operator
	Carpenter		Mixer Operator		Truck Driver
	Crane Operator		Motor Grader Operator		Utility Labor
	Flagger		Pipelayer		Work Zone Barricades
	Foreman		Roller Operator		
	Form Builder/Setter Struc				

Description of Work Performed						
Activity	1	2	3	4	5	
Job No.						Total
Category						
Function						
Name	Number of hours worked on each activity					

Activity	Description
1	
2	
3	
4	
5	

MEASUREMENTS AND CALCULATIONS							
DESCRIPTION	LOCATION	QTY	LENGTH	WIDTH	HEIGHT	CONV FACTOR	VOLUME OR AREA

ALBERTO MUNGUIA MIRELES

FOCAL POINTS

1	Weather statement	7	Unusual construction work conditions
2	Results of surveillance	8	Rework or field errors
3	Important verbal instructions received or given	9	Warranty work done
4	Names of official visitors and summary of discussion	10	Customer adjustments
5	Result of safety inspection	11	Estimating errors
6	Testing performed	12	Pay work item

Date: Construction's days remaining

EQ	QTY			CUT/		FILL/		Model or
	U	I	B	FROM	TO	FROM	TO	Notes
Compactors								
Compresors								
Cranes								
Excavators								
Graders								
Loader								
Man lift								
Trucks								
Tractors								
Scraper								
Mixer								

ALBERTO MUNGUIA MIRELES

Date:		Day	Night	S	M	T	W	T	F	S

Qty	Job Position	Qty	Job Position	Qty	Job Position
	Backhoe Operator		Front End Loader Operator		Scraper Operator
	Broom or Sweeper Op.		Hydraulic Excavator Op.		Serviceman
	Bulldozer Operator		Laborer		Steel Setter
	Concrete Finisher		Leadman		Surveyor
	Concrete Saw Operator		Mechanic		Tractor Operator
	Carpenter		Mixer Operator		Truck Driver
	Crane Operator		Motor Grader Operator		Utility Labor
	Flagger		Pipelayer		Work Zone Barricades
	Foreman		Roller Operator		
	Form Builder/Setter Struc				

Description of Work Performed						
Activity	1	2	3	4	5	
Job No.						Total
Category						
Function						
Name	Number of hours worked on each activity					

Activity	Description
1	
2	
3	
4	
5	

MEASUREMENTS AND CALCULATIONS							
DESCRIPTION	LOCATION	QTY	LENGTH	WIDTH	HEIGHT	CONV FACTOR	VOLUME OR AREA

ALBERTO MUNGUIA MIRELES

FOCAL POINTS

1	Weather statement	7	Unusual construction work conditions
2	Results of surveillance	8	Rework or field errors
3	Important verbal instructions received or given	9	Warranty work done
4	Names of official visitors and summary of discussion	10	Customer adjustments
5	Result of safety inspection	11	Estimating errors
6	Testing performed	12	Pay work item

Date: Construction's days remaining

EQ	QTY			CUT/		FILL/		Model or
	U	I	B	FROM	TO	FROM	TO	Notes
Compactors								
Compresors								
Cranes								
Excavators								
Graders								
Loader								
Man lift								
Trucks								
Tractors								
Scraper								
Mixer								

ALBERTO MUNGUIA MIRELES

Date:		Day	Night	S	M	T	W	T	F	S

Qty	Job Position	Qty	Job Position	Qty	Job Position
	Backhoe Operator		Front End Loader Operator		Scraper Operator
	Broom or Sweeper Op.		Hydraulic Excavator Op.		Serviceman
	Bulldozer Operator		Laborer		Steel Setter
	Concrete Finisher		Leadman		Surveyor
	Concrete Saw Operator		Mechanic		Tractor Operator
	Carpenter		Mixer Operator		Truck Driver
	Crane Operator		Motor Grader Operator		Utility Labor
	Flagger		Pipelayer		Work Zone Barricades
	Foreman		Roller Operator		
	Form Builder/Setter Struc				

Description of Work Performed						
Activity	1	2	3	4	5	
Job No.						**Total**
Category						
Function						
Name	Number of hours worked on each activity					

Activity	Description
1	
2	
3	
4	
5	

MEASUREMENTS AND CALCULATIONS							
DESCRIPTION	LOCATION	QTY	LENGTH	WIDTH	HEIGHT	CONV FACTOR	VOLUME OR AREA

ALBERTO MUNGUIA MIRELES

FOCAL POINTS

1	Weather statement	7	Unusual construction work conditions
2	Results of surveillance	8	Rework or field errors
3	Important verbal instructions received or given	9	Warranty work done
4	Names of official visitors and summary of discussion	10	Customer adjustments
5	Result of safety inspection	11	Estimating errors
6	Testing performed	12	Pay work item

Date: Construction's days remaining

EQ	QTY			CUT/		FILL/		Model or
	U	I	B	FROM	TO	FROM	TO	Notes
Compactors								
Compresors								
Cranes								
Excavators								
Graders								
Loader								
Man lift								
Trucks								
Tractors								
Scraper								
Mixer								

ALBERTO MUNGUIA MIRELES

Date:		Day	Night	S	M	T	W	T	F	S

Qty	Job Position	Qty	Job Position	Qty	Job Position
	Backhoe Operator		Front End Loader Operator		Scraper Operator
	Broom or Sweeper Op.		Hydraulic Excavator Op.		Serviceman
	Bulldozer Operator		Laborer		Steel Setter
	Concrete Finisher		Leadman		Surveyor
	Concrete Saw Operator		Mechanic		Tractor Operator
	Carpenter		Mixer Operator		Truck Driver
	Crane Operator		Motor Grader Operator		Utility Labor
	Flagger		Pipelayer		Work Zone Barricades
	Foreman		Roller Operator		
	Form Builder/Setter Struc				

Description of Work Performed						
Activity	**1**	**2**	**3**	**4**	**5**	
Job No.						**Total**
Category						
Function						
Name	Number of hours worked on each activity					

Activity	Description
1	
2	
3	
4	
5	

MEASUREMENTS AND CALCULATIONS							
DESCRIPTION	LOCATION	QTY	LENGTH	WIDTH	HEIGHT	CONV FACTOR	VOLUME OR AREA

ALBERTO MUNGUIA MIRELES

FOCAL POINTS

1	Weather statement	7	Unusual construction work conditions
2	Results of surveillance	8	Rework or field errors
3	Important verbal instructions received or given	9	Warranty work done
4	Names of official visitors and summary of discussion	10	Customer adjustments
5	Result of safety inspection	11	Estimating errors
6	Testing performed	12	Pay work item

Date: Construction's days remaining

EQ	QTY			CUT/		FILL/		Model or
	U	I	B	FROM	TO	FROM	TO	Notes
Compactors								
Compresors								
Cranes								
Excavators								
Graders								
Loader								
Man lift								
Trucks								
Tractors								
Scraper								
Mixer								

ALBERTO MUNGUIA MIRELES

Date:		Day	Night	S	M	T	W	T	F	S

Qty	Job Position	Qty	Job Position	Qty	Job Position
	Backhoe Operator		Front End Loader Operator		Scraper Operator
	Broom or Sweeper Op.		Hydraulic Excavator Op.		Serviceman
	Bulldozer Operator		Laborer		Steel Setter
	Concrete Finisher		Leadman		Surveyor
	Concrete Saw Operator		Mechanic		Tractor Operator
	Carpenter		Mixer Operator		Truck Driver
	Crane Operator		Motor Grader Operator		Utility Labor
	Flagger		Pipelayer		Work Zone Barricades
	Foreman		Roller Operator		
	Form Builder/Setter Struc				

Description of Work Performed							
Activity	1	2	3	4	5	Total
Job No.						
Category						
Function						
Name	Number of hours worked on each activity					

Activity	Description
1	
2	
3	
4	
5	

HIGHWAY CONSTRUCTION AND INSPECTION FIELDBOOK

MEASUREMENTS AND CALCULATIONS							
DESCRIPTION	LOCATION	QTY	LENGTH	WIDTH	HEIGHT	CONV FACTOR	VOLUME OR AREA

ALBERTO MUNGUIA MIRELES

FOCAL POINTS

1	Weather statement	7	Unusual construction work conditions
2	Results of surveillance	8	Rework or field errors
3	Important verbal instructions received or given	9	Warranty work done
4	Names of official visitors and summary of discussion	10	Customer adjustments
5	Result of safety inspection	11	Estimating errors
6	Testing performed	12	Pay work item

Date: Construction's days remaining

EQ	QTY			CUT/		FILL/		Model or
	U	I	B	FROM	TO	FROM	TO	Notes
Compactors								
Compresors								
Cranes								
Excavators								
Graders								
Loader								
Man lift								
Trucks								
Tractors								
Scraper								
Mixer								

ALBERTO MUNGUIA MIRELES

Date:		Day	Night	S	M	T	W	T	F	S

Qty	Job Position	Qty	Job Position	Qty	Job Position
	Backhoe Operator		Front End Loader Operator		Scraper Operator
	Broom or Sweeper Op.		Hydraulic Excavator Op.		Serviceman
	Bulldozer Operator		Laborer		Steel Setter
	Concrete Finisher		Leadman		Surveyor
	Concrete Saw Operator		Mechanic		Tractor Operator
	Carpenter		Mixer Operator		Truck Driver
	Crane Operator		Motor Grader Operator		Utility Labor
	Flagger		Pipelayer		Work Zone Barricades
	Foreman		Roller Operator		
	Form Builder/Setter Struc				

Description of Work Performed						
Activity	1	2	3	4	5	
Job No.						**Total**
Category						
Function						
Name	Number of hours worked on each activity					

Activity	Description
1	
2	
3	
4	
5	

MEASUREMENTS AND CALCULATIONS							
DESCRIPTION	LOCATION	QTY	LENGTH	WIDTH	HEIGHT	CONV FACTOR	VOLUME OR AREA

ALBERTO MUNGUIA MIRELES

FOCAL POINTS

1	Weather statement	7	Unusual construction work conditions
2	Results of surveillance	8	Rework or field errors
3	Important verbal instructions received or given	9	Warranty work done
4	Names of official visitors and summary of discussion	10	Customer adjustments
5	Result of safety inspection	11	Estimating errors
6	Testing performed	12	Pay work item

Date: Construction's days remaining

HIGHWAY CONSTRUCTION AND INSPECTION FIELDBOOK

EQ	QTY			CUT/		FILL/		Model or
	U	I	B	FROM	TO	FROM	TO	Notes
Compactors								
Compresors								
Cranes								
Excavators								
Graders								
Loader								
Man lift								
Trucks								
Tractors								
Scraper								
Mixer								

ALBERTO MUNGUIA MIRELES

Date:		Day	Night	S	M	T	W	T	F	S

Qty	Job Position	Qty	Job Position	Qty	Job Position
	Backhoe Operator		Front End Loader Operator		Scraper Operator
	Broom or Sweeper Op.		Hydraulic Excavator Op.		Serviceman
	Bulldozer Operator		Laborer		Steel Setter
	Concrete Finisher		Leadman		Surveyor
	Concrete Saw Operator		Mechanic		Tractor Operator
	Carpenter		Mixer Operator		Truck Driver
	Crane Operator		Motor Grader Operator		Utility Labor
	Flagger		Pipelayer		Work Zone Barricades
	Foreman		Roller Operator		
	Form Builder/Setter Struc				

Description of Work Performed						
Activity	1	2	3	4	5	
Job No.						Total
Category						
Function						
Name	Number of hours worked on each activity					

Activity	Description
1	
2	
3	
4	
5	

MEASUREMENTS AND CALCULATIONS							
DESCRIPTION	LOCATION	QTY	LENGTH	WIDTH	HEIGHT	CONV FACTOR	VOLUME OR AREA

ALBERTO MUNGUIA MIRELES

FOCAL POINTS

1	Weather statement	7	Unusual construction work conditions
2	Results of surveillance	8	Rework or field errors
3	Important verbal instructions received or given	9	Warranty work done
4	Names of official visitors and summary of discussion	10	Customer adjustments
5	Result of safety inspection	11	Estimating errors
6	Testing performed	12	Pay work item

Date: Construction's days remaining

EQ	QTY			CUT/		FILL/		Model or
	U	I	B	FROM	TO	FROM	TO	Notes
Compactors								
Compresors								
Cranes								
Excavators								
Graders								
Loader								
Man lift								
Trucks								
Tractors								
Scraper								
Mixer								

ALBERTO MUNGUIA MIRELES

Date:		Day	Night	S	M	T	W	T	F	S

Qty	Job Position	Qty	Job Position	Qty	Job Position
	Backhoe Operator		Front End Loader Operator		Scraper Operator
	Broom or Sweeper Op.		Hydraulic Excavator Op.		Serviceman
	Bulldozer Operator		Laborer		Steel Setter
	Concrete Finisher		Leadman		Surveyor
	Concrete Saw Operator		Mechanic		Tractor Operator
	Carpenter		Mixer Operator		Truck Driver
	Crane Operator		Motor Grader Operator		Utility Labor
	Flagger		Pipelayer		Work Zone Barricades
	Foreman		Roller Operator		
	Form Builder/Setter Struc				

Description of Work Performed						
Activity	**1**	**2**	**3**	**4**	**5**	
Job No.						Total
Category						
Function						
Name	Number of hours worked on each activity					

Activity	Description
1	
2	
3	
4	
5	

MEASUREMENTS AND CALCULATIONS							
DESCRIPTION	LOCATION	QTY	LENGTH	WIDTH	HEIGHT	CONV FACTOR	VOLUME OR AREA

ALBERTO MUNGUIA MIRELES

FOCAL POINTS

1	Weather statement	7	Unusual construction work conditions
2	Results of surveillance	8	Rework or field errors
3	Important verbal instructions received or given	9	Warranty work done
4	Names of official visitors and summary of discussion	10	Customer adjustments
5	Result of safety inspection	11	Estimating errors
6	Testing performed	12	Pay work item

Date: Construction's days remaining

EQ	QTY			CUT/		FILL/		Model or
	U	I	B	FROM	TO	FROM	TO	Notes
Compactors								
Compresors								
Cranes								
Excavators								
Graders								
Loader								
Man lift								
Trucks								
Tractors								
Scraper								
Mixer								

ALBERTO MUNGUIA MIRELES

Date:		Day	Night	S	M	T	W	T	F	S

Qty	Job Position	Qty	Job Position	Qty	Job Position
	Backhoe Operator		Front End Loader Operator		Scraper Operator
	Broom or Sweeper Op.		Hydraulic Excavator Op.		Serviceman
	Bulldozer Operator		Laborer		Steel Setter
	Concrete Finisher		Leadman		Surveyor
	Concrete Saw Operator		Mechanic		Tractor Operator
	Carpenter		Mixer Operator		Truck Driver
	Crane Operator		Motor Grader Operator		Utility Labor
	Flagger		Pipelayer		Work Zone Barricades
	Foreman		Roller Operator		
	Form Builder/Setter Struc				

Description of Work Performed						
Activity	1	2	3	4	5	
Job No.						Total
Category						
Function						
Name	Number of hours worked on each activity					

Activity	Description
1	
2	
3	
4	
5	

MEASUREMENTS AND CALCULATIONS							
DESCRIPTION	LOCATION	QTY	LENGTH	WIDTH	HEIGHT	CONV FACTOR	VOLUME OR AREA

ALBERTO MUNGUIA MIRELES

FOCAL POINTS

1	Weather statement	7	Unusual construction work conditions
2	Results of surveillance	8	Rework or field errors
3	Important verbal instructions received or given	9	Warranty work done
4	Names of official visitors and summary of discussion	10	Customer adjustments
5	Result of safety inspection	11	Estimating errors
6	Testing performed	12	Pay work item

Date: Construction's days remaining

EQ	QTY			CUT/		FILL/		Model or
	U	I	B	FROM	TO	FROM	TO	Notes
Compactors								
Compresors								
Cranes								
Excavators								
Graders								
Loader								
Man lift								
Trucks								
Tractors								
Scraper								
Mixer								

ALBERTO MUNGUIA MIRELES

Date:		Day	Night	S	M	T	W	T	F	S

Qty	Job Position	Qty	Job Position	Qty	Job Position
	Backhoe Operator		Front End Loader Operator		Scraper Operator
	Broom or Sweeper Op.		Hydraulic Excavator Op.		Serviceman
	Bulldozer Operator		Laborer		Steel Setter
	Concrete Finisher		Leadman		Surveyor
	Concrete Saw Operator		Mechanic		Tractor Operator
	Carpenter		Mixer Operator		Truck Driver
	Crane Operator		Motor Grader Operator		Utility Labor
	Flagger		Pipelayer		Work Zone Barricades
	Foreman		Roller Operator		
	Form Builder/Setter Struc				

Description of Work Performed						
Activity	1	2	3	4	5	
Job No.						Total
Category						
Function						
Name	Number of hours worked on each activity					

Activity	Description
1	
2	
3	
4	
5	

MEASUREMENTS AND CALCULATIONS							
DESCRIPTION	LOCATION	QTY	LENGTH	WIDTH	HEIGHT	CONV FACTOR	VOLUME OR AREA

FOCAL POINTS

1	Weather statement	7	Unusual construction work conditions
2	Results of surveillance	8	Rework or field errors
3	Important verbal instructions received or given	9	Warranty work done
4	Names of official visitors and summary of discussion	10	Customer adjustments
5	Result of safety inspection	11	Estimating errors
6	Testing performed	12	Pay work item

Date: Construction's days remaining

EQ	QTY			CUT/		FILL/		Model or
	U	I	B	FROM	TO	FROM	TO	Notes
Compactors								
Compresors								
Cranes								
Excavators								
Graders								
Loader								
Man lift								
Trucks								
Tractors								
Scraper								
Mixer								

ALBERTO MUNGUIA MIRELES

Date:		Day	Night	S	M	T	W	T	F	S

Qty	Job Position	Qty	Job Position	Qty	Job Position
	Backhoe Operator		Front End Loader Operator		Scraper Operator
	Broom or Sweeper Op.		Hydraulic Excavator Op.		Serviceman
	Bulldozer Operator		Laborer		Steel Setter
	Concrete Finisher		Leadman		Surveyor
	Concrete Saw Operator		Mechanic		Tractor Operator
	Carpenter		Mixer Operator		Truck Driver
	Crane Operator		Motor Grader Operator		Utility Labor
	Flagger		Pipelayer		Work Zone Barricades
	Foreman		Roller Operator		
	Form Builder/Setter Struc				

Description of Work Performed						
Activity	1	2	3	4	5	
Job No.						**Total**
Category						
Function						
Name	Number of hours worked on each activity					

Activity	Description
1	
2	
3	
4	
5	

MEASUREMENTS AND CALCULATIONS							
DESCRIPTION	LOCATION	QTY	LENGTH	WIDTH	HEIGHT	CONV FACTOR	VOLUME OR AREA

ALBERTO MUNGUIA MIRELES

FOCAL POINTS

1	Weather statement	7	Unusual construction work conditions
2	Results of surveillance	8	Rework or field errors
3	Important verbal instructions received or given	9	Warranty work done
4	Names of official visitors and summary of discussion	10	Customer adjustments
5	Result of safety inspection	11	Estimating errors
6	Testing performed	12	Pay work item

Date: Construction's days remaining

EQ	QTY			CUT/		FILL/		Model or
	U	I	B	FROM	TO	FROM	TO	Notes
Compactors								
Compresors								
Cranes								
Excavators								
Graders								
Loader								
Man lift								
Trucks								
Tractors								
Scraper								
Mixer								

ALBERTO MUNGUIA MIRELES

Date:		Day	Night	S	M	T	W	T	F	S

Qty	Job Position	Qty	Job Position	Qty	Job Position
	Backhoe Operator		Front End Loader Operator		Scraper Operator
	Broom or Sweeper Op.		Hydraulic Excavator Op.		Serviceman
	Bulldozer Operator		Laborer		Steel Setter
	Concrete Finisher		Leadman		Surveyor
	Concrete Saw Operator		Mechanic		Tractor Operator
	Carpenter		Mixer Operator		Truck Driver
	Crane Operator		Motor Grader Operator		Utility Labor
	Flagger		Pipelayer		Work Zone Barricades
	Foreman		Roller Operator		
	Form Builder/Setter Struc				

Description of Work Performed						
Activity	1	2	3	4	5	
Job No.						Total
Category						
Function						
Name	Number of hours worked on each activity					

Activity	Description
1	
2	
3	
4	
5	

MEASUREMENTS AND CALCULATIONS							
DESCRIPTION	LOCATION	QTY	LENGTH	WIDTH	HEIGHT	CONV FACTOR	VOLUME OR AREA

ALBERTO MUNGUIA MIRELES

FOCAL POINTS

1	Weather statement	7	Unusual construction work conditions
2	Results of surveillance	8	Rework or field errors
3	Important verbal instructions received or given	9	Warranty work done
4	Names of official visitors and summary of discussion	10	Customer adjustments
5	Result of safety inspection	11	Estimating errors
6	Testing performed	12	Pay work item

Date: Construction's days remaining

EQ	QTY			CUT/		FILL/		Model or
	U	I	B	FROM	TO	FROM	TO	Notes
Compactors								
Compresors								
Cranes								
Excavators								
Graders								
Loader								
Man lift								
Trucks								
Tractors								
Scraper								
Mixer								

ALBERTO MUNGUIA MIRELES

Date:		Day	Night	S	M	T	W	T	F	S

Qty	Job Position	Qty	Job Position	Qty	Job Position
	Backhoe Operator		Front End Loader Operator		Scraper Operator
	Broom or Sweeper Op.		Hydraulic Excavator Op.		Serviceman
	Bulldozer Operator		Laborer		Steel Setter
	Concrete Finisher		Leadman		Surveyor
	Concrete Saw Operator		Mechanic		Tractor Operator
	Carpenter		Mixer Operator		Truck Driver
	Crane Operator		Motor Grader Operator		Utility Labor
	Flagger		Pipelayer		Work Zone Barricades
	Foreman		Roller Operator		
	Form Builder/Setter Struc				

Description of Work Performed						
Activity	1	2	3	4	5	
Job No.						Total
Category						
Function						
Name	Number of hours worked on each activity					

Activity	Description
1	
2	
3	
4	
5	

MEASUREMENTS AND CALCULATIONS							
DESCRIPTION	LOCATION	QTY	LENGTH	WIDTH	HEIGHT	CONV FACTOR	VOLUME OR AREA

FOCAL POINTS

1	Weather statement	7	Unusual construction work conditions
2	Results of surveillance	8	Rework or field errors
3	Important verbal instructions received or given	9	Warranty work done
4	Names of official visitors and summary of discussion	10	Customer adjustments
5	Result of safety inspection	11	Estimating errors
6	Testing performed	12	Pay work item

Date: Construction's days remaining

EQ	QTY			CUT/		FILL/		Model or
	U	I	B	FROM	TO	FROM	TO	Notes
Compactors								
Compresors								
Cranes								
Excavators								
Graders								
Loader								
Man lift								
Trucks								
Tractors								
Scraper								
Mixer								

ALBERTO MUNGUIA MIRELES

Date:		Day	Night	S	M	T	W	T	F	S

Qty	Job Position	Qty	Job Position	Qty	Job Position
	Backhoe Operator		Front End Loader Operator		Scraper Operator
	Broom or Sweeper Op.		Hydraulic Excavator Op.		Serviceman
	Bulldozer Operator		Laborer		Steel Setter
	Concrete Finisher		Leadman		Surveyor
	Concrete Saw Operator		Mechanic		Tractor Operator
	Carpenter		Mixer Operator		Truck Driver
	Crane Operator		Motor Grader Operator		Utility Labor
	Flagger		Pipelayer		Work Zone Barricades
	Foreman		Roller Operator		
	Form Builder/Setter Struc				

Description of Work Performed						
Activity	**1**	**2**	**3**	**4**	**5**	
Job No.						**Total**
Category						
Function						
Name	Number of hours worked on each activity					

Activity	Description
1	
2	
3	
4	
5	

MEASUREMENTS AND CALCULATIONS							
DESCRIPTION	LOCATION	QTY	LENGTH	WIDTH	HEIGHT	CONV FACTOR	VOLUME OR AREA

ALBERTO MUNGUIA MIRELES

FOCAL POINTS

1	Weather statement	7	Unusual construction work conditions
2	Results of surveillance	8	Rework or field errors
3	Important verbal instructions received or given	9	Warranty work done
4	Names of official visitors and summary of discussion	10	Customer adjustments
5	Result of safety inspection	11	Estimating errors
6	Testing performed	12	Pay work item

Date: Construction's days remaining

EQ	QTY			CUT/		FILL/		Model or
	U	I	B	FROM	TO	FROM	TO	Notes
Compactors								
Compresors								
Cranes								
Excavators								
Graders								
Loader								
Man lift								
Trucks								
Tractors								
Scraper								
Mixer								

ALBERTO MUNGUIA MIRELES

Date:		Day	Night	S	M	T	W	T	F	S

Qty	Job Position	Qty	Job Position	Qty	Job Position
	Backhoe Operator		Front End Loader Operator		Scraper Operator
	Broom or Sweeper Op.		Hydraulic Excavator Op.		Serviceman
	Bulldozer Operator		Laborer		Steel Setter
	Concrete Finisher		Leadman		Surveyor
	Concrete Saw Operator		Mechanic		Tractor Operator
	Carpenter		Mixer Operator		Truck Driver
	Crane Operator		Motor Grader Operator		Utility Labor
	Flagger		Pipelayer		Work Zone Barricades
	Foreman		Roller Operator		
	Form Builder/Setter Struc				

Description of Work Performed						
Activity	1	2	3	4	5	
Job No.						Total
Category						
Function						
Name	Number of hours worked on each activity					

Activity	Description
1	
2	
3	
4	
5	

MEASUREMENTS AND CALCULATIONS							
DESCRIPTION	LOCATION	QTY	LENGTH	WIDTH	HEIGHT	CONV FACTOR	VOLUME OR AREA

ALBERTO MUNGUIA MIRELES

FOCAL POINTS

1	Weather statement	7	Unusual construction work conditions
2	Results of surveillance	8	Rework or field errors
3	Important verbal instructions received or given	9	Warranty work done
4	Names of official visitors and summary of discussion	10	Customer adjustments
5	Result of safety inspection	11	Estimating errors
6	Testing performed	12	Pay work item

Date: Construction's days remaining

EQ	QTY			CUT/		FILL/		Model or
	U	I	B	FROM	TO	FROM	TO	Notes
Compactors								
Compresors								
Cranes								
Excavators								
Graders								
Loader								
Man lift								
Trucks								
Tractors								
Scraper								
Mixer								

ALBERTO MUNGUIA MIRELES

Date:		Day	Night	S	M	T	W	T	F	S

Qty	Job Position	Qty	Job Position	Qty	Job Position
	Backhoe Operator		Front End Loader Operator		Scraper Operator
	Broom or Sweeper Op.		Hydraulic Excavator Op.		Serviceman
	Bulldozer Operator		Laborer		Steel Setter
	Concrete Finisher		Leadman		Surveyor
	Concrete Saw Operator		Mechanic		Tractor Operator
	Carpenter		Mixer Operator		Truck Driver
	Crane Operator		Motor Grader Operator		Utility Labor
	Flagger		Pipelayer		Work Zone Barricades
	Foreman		Roller Operator		
	Form Builder/Setter Struc				

Description of Work Performed						
Activity	1	2	3	4	5	
Job No.						Total
Category						
Function						
Name	Number of hours worked on each activity					

Activity	Description
1	
2	
3	
4	
5	

MEASUREMENTS AND CALCULATIONS							
DESCRIPTION	LOCATION	QTY	LENGTH	WIDTH	HEIGHT	CONV FACTOR	VOLUME OR AREA

ALBERTO MUNGUIA MIRELES

FOCAL POINTS

1	Weather statement	7	Unusual construction work conditions
2	Results of surveillance	8	Rework or field errors
3	Important verbal instructions received or given	9	Warranty work done
4	Names of official visitors and summary of discussion	10	Customer adjustments
5	Result of safety inspection	11	Estimating errors
6	Testing performed	12	Pay work item

Date: Construction's days remaining

EQ	QTY			CUT/		FILL/		Model or
	U	I	B	FROM	TO	FROM	TO	Notes
Compactors								
Compresors								
Cranes								
Excavators								
Graders								
Loader								
Man lift								
Trucks								
Tractors								
Scraper								
Mixer								

ALBERTO MUNGUIA MIRELES

Date:		Day	Night	S	M	T	W	T	F	S

Qty	Job Position	Qty	Job Position	Qty	Job Position
	Backhoe Operator		Front End Loader Operator		Scraper Operator
	Broom or Sweeper Op.		Hydraulic Excavator Op.		Serviceman
	Bulldozer Operator		Laborer		Steel Setter
	Concrete Finisher		Leadman		Surveyor
	Concrete Saw Operator		Mechanic		Tractor Operator
	Carpenter		Mixer Operator		Truck Driver
	Crane Operator		Motor Grader Operator		Utility Labor
	Flagger		Pipelayer		Work Zone Barricades
	Foreman		Roller Operator		
	Form Builder/Setter Struc				

Description of Work Performed						
Activity	1	2	3	4	5	
Job No.						**Total**
Category						
Function						
Name	Number of hours worked on each activity					

Activity	Description
1	
2	
3	
4	
5	

MEASUREMENTS AND CALCULATIONS							
DESCRIPTION	LOCATION	QTY	LENGTH	WIDTH	HEIGHT	CONV FACTOR	VOLUME OR AREA

ALBERTO MUNGUIA MIRELES

FOCAL POINTS

1	Weather statement	7	Unusual construction work conditions
2	Results of surveillance	8	Rework or field errors
3	Important verbal instructions received or given	9	Warranty work done
4	Names of official visitors and summary of discussion	10	Customer adjustments
5	Result of safety inspection	11	Estimating errors
6	Testing performed	12	Pay work item

Date: Construction's days remaining

EQ	QTY			CUT/		FILL/		Model or
	U	I	B	FROM	TO	FROM	TO	Notes
Compactors								
Compresors								
Cranes								
Excavators								
Graders								
Loader								
Man lift								
Trucks								
Tractors								
Scraper								
Mixer								

ALBERTO MUNGUIA MIRELES

Date:		Day	Night	S	M	T	W	T	F	S

Qty	Job Position	Qty	Job Position	Qty	Job Position
	Backhoe Operator		Front End Loader Operator		Scraper Operator
	Broom or Sweeper Op.		Hydraulic Excavator Op.		Serviceman
	Bulldozer Operator		Laborer		Steel Setter
	Concrete Finisher		Leadman		Surveyor
	Concrete Saw Operator		Mechanic		Tractor Operator
	Carpenter		Mixer Operator		Truck Driver
	Crane Operator		Motor Grader Operator		Utility Labor
	Flagger		Pipelayer		Work Zone Barricades
	Foreman		Roller Operator		
	Form Builder/Setter Struc				

Description of Work Performed						
Activity	1	2	3	4	5	
Job No.						Total
Category						
Function						
Name	Number of hours worked on each activity					

Activity	Description
1	
2	
3	
4	
5	

MEASUREMENTS AND CALCULATIONS							
DESCRIPTION	LOCATION	QTY	LENGTH	WIDTH	HEIGHT	CONV FACTOR	VOLUME OR AREA

ALBERTO MUNGUIA MIRELES

FOCAL POINTS

1	Weather statement	7	Unusual construction work conditions
2	Results of surveillance	8	Rework or field errors
3	Important verbal instructions received or given	9	Warranty work done
4	Names of official visitors and summary of discussion	10	Customer adjustments
5	Result of safety inspection	11	Estimating errors
6	Testing performed	12	Pay work item

Date: Construction's days remaining

HIGHWAY CONSTRUCTION AND INSPECTION FIELDBOOK

EQ	QTY			CUT/		FILL/		Model or
	U	I	B	FROM	TO	FROM	TO	Notes
Compactors								
Compresors								
Cranes								
Excavators								
Graders								
Loader								
Man lift								
Trucks								
Tractors								
Scraper								
Mixer								

ALBERTO MUNGUIA MIRELES

Date:		Day	Night	S	M	T	W	T	F	S

Qty	Job Position	Qty	Job Position	Qty	Job Position
	Backhoe Operator		Front End Loader Operator		Scraper Operator
	Broom or Sweeper Op.		Hydraulic Excavator Op.		Serviceman
	Bulldozer Operator		Laborer		Steel Setter
	Concrete Finisher		Leadman		Surveyor
	Concrete Saw Operator		Mechanic		Tractor Operator
	Carpenter		Mixer Operator		Truck Driver
	Crane Operator		Motor Grader Operator		Utility Labor
	Flagger		Pipelayer		Work Zone Barricades
	Foreman		Roller Operator		
	Form Builder/Setter Struc				

Description of Work Performed						
Activity	1	2	3	4	5	
Job No.						**Total**
Category						
Function						
Name	Number of hours worked on each activity					

Activity	Description
1	
2	
3	
4	
5	

MEASUREMENTS AND CALCULATIONS							
DESCRIPTION	LOCATION	QTY	LENGTH	WIDTH	HEIGHT	CONV FACTOR	VOLUME OR AREA

ALBERTO MUNGUIA MIRELES

FOCAL POINTS

1	Weather statement	7	Unusual construction work conditions
2	Results of surveillance	8	Rework or field errors
3	Important verbal instructions received or given	9	Warranty work done
4	Names of official visitors and summary of discussion	10	Customer adjustments
5	Result of safety inspection	11	Estimating errors
6	Testing performed	12	Pay work item

Date: Construction's days remaining

EQ	QTY			CUT/		FILL/		Model or
	U	I	B	FROM	TO	FROM	TO	Notes
Compactors								
Compresors								
Cranes								
Excavators								
Graders								
Loader								
Man lift								
Trucks								
Tractors								
Scraper								
Mixer								

ALBERTO MUNGUIA MIRELES

Date:		Day	Night	S	M	T	W	T	F	S

Qty	Job Position	Qty	Job Position	Qty	Job Position
	Backhoe Operator		Front End Loader Operator		Scraper Operator
	Broom or Sweeper Op.		Hydraulic Excavator Op.		Serviceman
	Bulldozer Operator		Laborer		Steel Setter
	Concrete Finisher		Leadman		Surveyor
	Concrete Saw Operator		Mechanic		Tractor Operator
	Carpenter		Mixer Operator		Truck Driver
	Crane Operator		Motor Grader Operator		Utility Labor
	Flagger		Pipelayer		Work Zone Barricades
	Foreman		Roller Operator		
	Form Builder/Setter Struc				

Description of Work Performed						
Activity	1	2	3	4	5	
Job No.						Total
Category						
Function						
Name	Number of hours worked on each activity					

Activity	Description
1	
2	
3	
4	
5	

MEASUREMENTS AND CALCULATIONS							
DESCRIPTION	LOCATION	QTY	LENGTH	WIDTH	HEIGHT	CONV FACTOR	VOLUME OR AREA

FOCAL POINTS

1	Weather statement	7	Unusual construction work conditions
2	Results of surveillance	8	Rework or field errors
3	Important verbal instructions received or given	9	Warranty work done
4	Names of official visitors and summary of discussion	10	Customer adjustments
5	Result of safety inspection	11	Estimating errors
6	Testing performed	12	Pay work item

Date: Construction's days remaining

EQ	QTY			CUT/		FILL/		Model or
	U	I	B	FROM	TO	FROM	TO	Notes
Compactors								
Compresors								
Cranes								
Excavators								
Graders								
Loader								
Man lift								
Trucks								
Tractors								
Scraper								
Mixer								

ALBERTO MUNGUIA MIRELES

Date:		Day	Night	S	M	T	W	T	F	S

Qty	Job Position	Qty	Job Position	Qty	Job Position
	Backhoe Operator		Front End Loader Operator		Scraper Operator
	Broom or Sweeper Op.		Hydraulic Excavator Op.		Serviceman
	Bulldozer Operator		Laborer		Steel Setter
	Concrete Finisher		Leadman		Surveyor
	Concrete Saw Operator		Mechanic		Tractor Operator
	Carpenter		Mixer Operator		Truck Driver
	Crane Operator		Motor Grader Operator		Utility Labor
	Flagger		Pipelayer		Work Zone Barricades
	Foreman		Roller Operator		
	Form Builder/Setter Struc				

Description of Work Performed						
Activity	1	2	3	4	5	
Job No.						Total
Category						
Function						
Name	Number of hours worked on each activity					

Activity	Description
1	
2	
3	
4	
5	

MEASUREMENTS AND CALCULATIONS							
DESCRIPTION	LOCATION	QTY	LENGTH	WIDTH	HEIGHT	CONV FACTOR	VOLUME OR AREA

ALBERTO MUNGUIA MIRELES

FOCAL POINTS

1	Weather statement	7	Unusual construction work conditions
2	Results of surveillance	8	Rework or field errors
3	Important verbal instructions received or given	9	Warranty work done
4	Names of official visitors and summary of discussion	10	Customer adjustments
5	Result of safety inspection	11	Estimating errors
6	Testing performed	12	Pay work item

Date: Construction's days remaining

EQ	QTY			CUT/		FILL/		Model or
	U	I	B	FROM	TO	FROM	TO	Notes
Compactors								
Compresors								
Cranes								
Excavators								
Graders								
Loader								
Man lift								
Trucks								
Tractors								
Scraper								
Mixer								

ALBERTO MUNGUIA MIRELES

Date:		Day	Night	S	M	T	W	T	F	S

Qty	Job Position	Qty	Job Position	Qty	Job Position
	Backhoe Operator		Front End Loader Operator		Scraper Operator
	Broom or Sweeper Op.		Hydraulic Excavator Op.		Serviceman
	Bulldozer Operator		Laborer		Steel Setter
	Concrete Finisher		Leadman		Surveyor
	Concrete Saw Operator		Mechanic		Tractor Operator
	Carpenter		Mixer Operator		Truck Driver
	Crane Operator		Motor Grader Operator		Utility Labor
	Flagger		Pipelayer		Work Zone Barricades
	Foreman		Roller Operator		
	Form Builder/Setter Struc				

Description of Work Performed						
Activity	1	2	3	4	5	
Job No.						**Total**
Category						
Function						
Name	Number of hours worked on each activity					

Activity	Description
1	
2	
3	
4	
5	

MEASUREMENTS AND CALCULATIONS							
DESCRIPTION	LOCATION	QTY	LENGTH	WIDTH	HEIGHT	CONV FACTOR	VOLUME OR AREA

ALBERTO MUNGUIA MIRELES

FOCAL POINTS

1	Weather statement	7	Unusual construction work conditions
2	Results of surveillance	8	Rework or field errors
3	Important verbal instructions received or given	9	Warranty work done
4	Names of official visitors and summary of discussion	10	Customer adjustments
5	Result of safety inspection	11	Estimating errors
6	Testing performed	12	Pay work item

Date: Construction's days remaining

HIGHWAY CONSTRUCTION AND INSPECTION FIELDBOOK

EQ	QTY			CUT/		FILL/		Model or
	U	I	B	FROM	TO	FROM	TO	Notes
Compactors								
Compresors								
Cranes								
Excavators								
Graders								
Loader								
Man lift								
Trucks								
Tractors								
Scraper								
Mixer								

ALBERTO MUNGUIA MIRELES

Date:		Day	Night	S	M	T	W	T	F	S

Qty	Job Position	Qty	Job Position	Qty	Job Position
	Backhoe Operator		Front End Loader Operator		Scraper Operator
	Broom or Sweeper Op.		Hydraulic Excavator Op.		Serviceman
	Bulldozer Operator		Laborer		Steel Setter
	Concrete Finisher		Leadman		Surveyor
	Concrete Saw Operator		Mechanic		Tractor Operator
	Carpenter		Mixer Operator		Truck Driver
	Crane Operator		Motor Grader Operator		Utility Labor
	Flagger		Pipelayer		Work Zone Barricades
	Foreman		Roller Operator		
	Form Builder/Setter Struc				

Description of Work Performed						
Activity	**1**	**2**	**3**	**4**	**5**	
Job No.						**Total**
Category						
Function						
Name	Number of hours worked on each activity					

Activity	Description
1	
2	
3	
4	
5	

MEASUREMENTS AND CALCULATIONS							
DESCRIPTION	LOCATION	QTY	LENGTH	WIDTH	HEIGHT	CONV FACTOR	VOLUME OR AREA

ALBERTO MUNGUIA MIRELES

FOCAL POINTS

1	Weather statement	7	Unusual construction work conditions
2	Results of surveillance	8	Rework or field errors
3	Important verbal instructions received or given	9	Warranty work done
4	Names of official visitors and summary of discussion	10	Customer adjustments
5	Result of safety inspection	11	Estimating errors
6	Testing performed	12	Pay work item

Date: Construction's days remaining

124

HIGHWAY CONSTRUCTION AND INSPECTION FIELDBOOK

EQ	QTY			CUT/		FILL/		Model or
	U	I	B	FROM	TO	FROM	TO	Notes
Compactors								
Compresors								
Cranes								
Excavators								
Graders								
Loader								
Man lift								
Trucks								
Tractors								
Scraper								
Mixer								

ALBERTO MUNGUIA MIRELES

Date:		Day	Night	S	M	T	W	T	F	S

Qty	Job Position	Qty	Job Position	Qty	Job Position
	Backhoe Operator		Front End Loader Operator		Scraper Operator
	Broom or Sweeper Op.		Hydraulic Excavator Op.		Serviceman
	Bulldozer Operator		Laborer		Steel Setter
	Concrete Finisher		Leadman		Surveyor
	Concrete Saw Operator		Mechanic		Tractor Operator
	Carpenter		Mixer Operator		Truck Driver
	Crane Operator		Motor Grader Operator		Utility Labor
	Flagger		Pipelayer		Work Zone Barricades
	Foreman		Roller Operator		
	Form Builder/Setter Struc				

Description of Work Performed						
Activity	1	2	3	4	5	Total
Job No.						
Category						
Function						
Name	Number of hours worked on each activity					

Activity	Description
1	
2	
3	
4	
5	

MEASUREMENTS AND CALCULATIONS							
DESCRIPTION	LOCATION	QTY	LENGTH	WIDTH	HEIGHT	CONV FACTOR	VOLUME OR AREA

ALBERTO MUNGUIA MIRELES

FOCAL POINTS

1	Weather statement	7	Unusual construction work conditions
2	Results of surveillance	8	Rework or field errors
3	Important verbal instructions received or given	9	Warranty work done
4	Names of official visitors and summary of discussion	10	Customer adjustments
5	Result of safety inspection	11	Estimating errors
6	Testing performed	12	Pay work item

Date: Construction's days remaining

EQ	QTY			CUT/		FILL/		Model or
	U	I	B	FROM	TO	FROM	TO	Notes
Compactors								
Compresors								
Cranes								
Excavators								
Graders								
Loader								
Man lift								
Trucks								
Tractors								
Scraper								
Mixer								

ALBERTO MUNGUIA MIRELES

Date:		Day	Night	S	M	T	W	T	F	S

Qty	Job Position	Qty	Job Position	Qty	Job Position
	Backhoe Operator		Front End Loader Operator		Scraper Operator
	Broom or Sweeper Op.		Hydraulic Excavator Op.		Serviceman
	Bulldozer Operator		Laborer		Steel Setter
	Concrete Finisher		Leadman		Surveyor
	Concrete Saw Operator		Mechanic		Tractor Operator
	Carpenter		Mixer Operator		Truck Driver
	Crane Operator		Motor Grader Operator		Utility Labor
	Flagger		Pipelayer		Work Zone Barricades
	Foreman		Roller Operator		
	Form Builder/Setter Struc				

Description of Work Performed						
Activity	1	2	3	4	5	
Job No.						Total
Category						
Function						
Name	Number of hours worked on each activity					

Activity	Description
1	
2	
3	
4	
5	

MEASUREMENTS AND CALCULATIONS							
DESCRIPTION	LOCATION	QTY	LENGTH	WIDTH	HEIGHT	CONV FACTOR	VOLUME OR AREA

FOCAL POINTS

1	Weather statement	7	Unusual construction work conditions
2	Results of surveillance	8	Rework or field errors
3	Important verbal instructions received or given	9	Warranty work done
4	Names of official visitors and summary of discussion	10	Customer adjustments
5	Result of safety inspection	11	Estimating errors
6	Testing performed	12	Pay work item

Date: Construction's days remaining

EQ	QTY			CUT/		FILL/		Model or
	U	I	B	FROM	TO	FROM	TO	Notes
Compactors								
Compresors								
Cranes								
Excavators								
Graders								
Loader								
Man lift								
Trucks								
Tractors								
Scraper								
Mixer								

ALBERTO MUNGUIA MIRELES

Date:		Day	Night	S	M	T	W	T	F	S

Qty	Job Position	Qty	Job Position	Qty	Job Position
	Backhoe Operator		Front End Loader Operator		Scraper Operator
	Broom or Sweeper Op.		Hydraulic Excavator Op.		Serviceman
	Bulldozer Operator		Laborer		Steel Setter
	Concrete Finisher		Leadman		Surveyor
	Concrete Saw Operator		Mechanic		Tractor Operator
	Carpenter		Mixer Operator		Truck Driver
	Crane Operator		Motor Grader Operator		Utility Labor
	Flagger		Pipelayer		Work Zone Barricades
	Foreman		Roller Operator		
	Form Builder/Setter Struc				

Description of Work Performed						
Activity	1	2	3	4	5	
Job No.						**Total**
Category						
Function						
Name	Number of hours worked on each activity					

Activity	Description
1	
2	
3	
4	
5	

MEASUREMENTS AND CALCULATIONS							
DESCRIPTION	LOCATION	QTY	LENGTH	WIDTH	HEIGHT	CONV FACTOR	VOLUME OR AREA

ALBERTO MUNGUIA MIRELES

FOCAL POINTS

1	Weather statement	7	Unusual construction work conditions
2	Results of surveillance	8	Rework or field errors
3	Important verbal instructions received or given	9	Warranty work done
4	Names of official visitors and summary of discussion	10	Customer adjustments
5	Result of safety inspection	11	Estimating errors
6	Testing performed	12	Pay work item

Date: Construction's days remaining

EQ	QTY			CUT/		FILL/		Model or
	U	I	B	FROM	TO	FROM	TO	Notes
Compactors								
Compresors								
Cranes								
Excavators								
Graders								
Loader								
Man lift								
Trucks								
Tractors								
Scraper								
Mixer								

ALBERTO MUNGUIA MIRELES

Date:		Day	Night	S	M	T	W	T	F	S

Qty	Job Position	Qty	Job Position	Qty	Job Position
	Backhoe Operator		Front End Loader Operator		Scraper Operator
	Broom or Sweeper Op.		Hydraulic Excavator Op.		Serviceman
	Bulldozer Operator		Laborer		Steel Setter
	Concrete Finisher		Leadman		Surveyor
	Concrete Saw Operator		Mechanic		Tractor Operator
	Carpenter		Mixer Operator		Truck Driver
	Crane Operator		Motor Grader Operator		Utility Labor
	Flagger		Pipelayer		Work Zone Barricades
	Foreman		Roller Operator		
	Form Builder/Setter Struc				

Description of Work Performed						
Activity	1	2	3	4	5	
Job No.						**Total**
Category						
Function						
Name	Number of hours worked on each activity					

Activity	Description
1	
2	
3	
4	
5	

MEASUREMENTS AND CALCULATIONS							
DESCRIPTION	LOCATION	QTY	LENGTH	WIDTH	HEIGHT	CONV FACTOR	VOLUME OR AREA

ALBERTO MUNGUIA MIRELES

FOCAL POINTS

1	Weather statement	7	Unusual construction work conditions
2	Results of surveillance	8	Rework or field errors
3	Important verbal instructions received or given	9	Warranty work done
4	Names of official visitors and summary of discussion	10	Customer adjustments
5	Result of safety inspection	11	Estimating errors
6	Testing performed	12	Pay work item

Date: Construction's days remaining

HIGHWAY CONSTRUCTION AND INSPECTION FIELDBOOK

EQ	QTY			CUT/		FILL/		Model or
	U	I	B	FROM	TO	FROM	TO	Notes
Compactors								
Compresors								
Cranes								
Excavators								
Graders								
Loader								
Man lift								
Trucks								
Tractors								
Scraper								
Mixer								

ALBERTO MUNGUIA MIRELES

Date:		Day	Night	S	M	T	W	T	F	S

Qty	Job Position	Qty	Job Position	Qty	Job Position
	Backhoe Operator		Front End Loader Operator		Scraper Operator
	Broom or Sweeper Op.		Hydraulic Excavator Op.		Serviceman
	Bulldozer Operator		Laborer		Steel Setter
	Concrete Finisher		Leadman		Surveyor
	Concrete Saw Operator		Mechanic		Tractor Operator
	Carpenter		Mixer Operator		Truck Driver
	Crane Operator		Motor Grader Operator		Utility Labor
	Flagger		Pipelayer		Work Zone Barricades
	Foreman		Roller Operator		
	Form Builder/Setter Struc				

Description of Work Performed						
Activity	1	2	3	4	5	
Job No.						**Total**
Category						
Function						
Name	Number of hours worked on each activity					

Activity	Description
1	
2	
3	
4	
5	

MEASUREMENTS AND CALCULATIONS							
DESCRIPTION	LOCATION	QTY	LENGTH	WIDTH	HEIGHT	CONV FACTOR	VOLUME OR AREA

ALBERTO MUNGUIA MIRELES

FOCAL POINTS

1	Weather statement	7	Unusual construction work conditions
2	Results of surveillance	8	Rework or field errors
3	Important verbal instructions received or given	9	Warranty work done
4	Names of official visitors and summary of discussion	10	Customer adjustments
5	Result of safety inspection	11	Estimating errors
6	Testing performed	12	Pay work item

Date: Construction's days remaining

EQ	QTY			CUT/		FILL/		Model or
	U	I	B	FROM	TO	FROM	TO	Notes
Compactors								
Compresors								
Cranes								
Excavators								
Graders								
Loader								
Man lift								
Trucks								
Tractors								
Scraper								
Mixer								

ALBERTO MUNGUIA MIRELES

Date:		Day	Night	S	M	T	W	T	F	S

Qty	Job Position	Qty	Job Position	Qty	Job Position
	Backhoe Operator		Front End Loader Operator		Scraper Operator
	Broom or Sweeper Op.		Hydraulic Excavator Op.		Serviceman
	Bulldozer Operator		Laborer		Steel Setter
	Concrete Finisher		Leadman		Surveyor
	Concrete Saw Operator		Mechanic		Tractor Operator
	Carpenter		Mixer Operator		Truck Driver
	Crane Operator		Motor Grader Operator		Utility Labor
	Flagger		Pipelayer		Work Zone Barricades
	Foreman		Roller Operator		
	Form Builder/Setter Struc				

Description of Work Performed						
Activity	1	2	3	4	5	
Job No.						Total
Category						
Function						
Name	Number of hours worked on each activity					

Activity	Description
1	
2	
3	
4	
5	

MEASUREMENTS AND CALCULATIONS							
DESCRIPTION	LOCATION	QTY	LENGTH	WIDTH	HEIGHT	CONV FACTOR	VOLUME OR AREA

ALBERTO MUNGUIA MIRELES

FOCAL POINTS

1	Weather statement	7	Unusual construction work conditions
2	Results of surveillance	8	Rework or field errors
3	Important verbal instructions received or given	9	Warranty work done
4	Names of official visitors and summary of discussion	10	Customer adjustments
5	Result of safety inspection	11	Estimating errors
6	Testing performed	12	Pay work item

Date: Construction's days remaining

EQ	QTY			CUT/		FILL/		Model or
	U	I	B	FROM	TO	FROM	TO	Notes
Compactors								
Compresors								
Cranes								
Excavators								
Graders								
Loader								
Man lift								
Trucks								
Tractors								
Scraper								
Mixer								

ALBERTO MUNGUIA MIRELES

Date:		Day	Night	S	M	T	W	T	F	S

Qty	Job Position	Qty	Job Position	Qty	Job Position
	Backhoe Operator		Front End Loader Operator		Scraper Operator
	Broom or Sweeper Op.		Hydraulic Excavator Op.		Serviceman
	Bulldozer Operator		Laborer		Steel Setter
	Concrete Finisher		Leadman		Surveyor
	Concrete Saw Operator		Mechanic		Tractor Operator
	Carpenter		Mixer Operator		Truck Driver
	Crane Operator		Motor Grader Operator		Utility Labor
	Flagger		Pipelayer		Work Zone Barricades
	Foreman		Roller Operator		
	Form Builder/Setter Struc				

Description of Work Performed						
Activity	1	2	3	4	5	
Job No.						Total
Category						
Function						
Name	Number of hours worked on each activity					

Activity	Description
1	
2	
3	
4	
5	

MEASUREMENTS AND CALCULATIONS							
DESCRIPTION	LOCATION	QTY	LENGTH	WIDTH	HEIGHT	CONV FACTOR	VOLUME OR AREA

ALBERTO MUNGUIA MIRELES

FOCAL POINTS

1	Weather statement	7	Unusual construction work conditions
2	Results of surveillance	8	Rework or field errors
3	Important verbal instructions received or given	9	Warranty work done
4	Names of official visitors and summary of discussion	10	Customer adjustments
5	Result of safety inspection	11	Estimating errors
6	Testing performed	12	Pay work item

Date: Construction's days remaining

EQ	QTY			CUT/		FILL/		Model or
	U	I	B	FROM	TO	FROM	TO	Notes
Compactors								
Compresors								
Cranes								
Excavators								
Graders								
Loader								
Man lift								
Trucks								
Tractors								
Scraper								
Mixer								

Date:		Day	Night	S	M	T	W	T	F	S

Qty	Job Position	Qty	Job Position	Qty	Job Position
	Backhoe Operator		Front End Loader Operator		Scraper Operator
	Broom or Sweeper Op.		Hydraulic Excavator Op.		Serviceman
	Bulldozer Operator		Laborer		Steel Setter
	Concrete Finisher		Leadman		Surveyor
	Concrete Saw Operator		Mechanic		Tractor Operator
	Carpenter		Mixer Operator		Truck Driver
	Crane Operator		Motor Grader Operator		Utility Labor
	Flagger		Pipelayer		Work Zone Barricades
	Foreman		Roller Operator		
	Form Builder/Setter Struc				

Description of Work Performed						
Activity	1	2	3	4	5	Total
Job No.						
Category						
Function						
Name	Number of hours worked on each activity					

Activity	Description
1	
2	
3	
4	
5	

MEASUREMENTS AND CALCULATIONS							
DESCRIPTION	LOCATION	QTY	LENGTH	WIDTH	HEIGHT	CONV FACTOR	VOLUME OR AREA

FOCAL POINTS

1	Weather statement	7	Unusual construction work conditions
2	Results of surveillance	8	Rework or field errors
3	Important verbal instructions received or given	9	Warranty work done
4	Names of official visitors and summary of discussion	10	Customer adjustments
5	Result of safety inspection	11	Estimating errors
6	Testing performed	12	Pay work item

Date: Construction's days remaining

Example 2

How to Fill Out the Log
for Lost Days

LOG FOR LOST DAYS

DATE			DUE TO		RAIN		
MM	DD	YY	RAIN/MUD	OTHER	INCHES	FROM	TO
2	2	12	I		1"	16:00	18:00
2	3	12	I		I	18:00	12:00
2	4	12	I		0.5	9:30	11:00
3	7	12	I		1.5	13:00	15:00
5	26	12		I	Contractor	Safety meeting	
7	5	12	I		1.5	10:15	12:00
	Subtotal		5	I			

Section 2

Filling Out the Log for Lost Days

DATE			DUE TO		RAIN		
MM	DD	YY	RAIN/MUD	OTHER	INCHES	FROM	TO
SUBTOTAL							

Lesson 1

Work As a Team

In a world rich in technology, social media, and many electronic ways of communication within projects, we often forget that we are dealing with people that should share the same goal.

Designers, local governments, contractors, and inspectors need to be aware that the only effective way that a contractor can make money and complete a project on time is by performing the work right on *the first try*. The only way an inspector can authorize payment to a contractor is when he or she is able to demonstrate that the work done by the contractor meets quality specifications. The only way that the owner of the project can accept the work is when it has been demonstrated that the project will provide the service for which it was undertaken—according to design.

It is, therefore, beneficial for everybody to perform the job as a team, collaborating to solve issues ahead of time before they become problems. Sometimes we forget that the construction phase of a project is the last phase of a *design* process. And the design is not only what is on the CAD drawings. It is what gets built with people that operate equipment, coordinate the delivery of material, and, most importantly, interact.

Therefore, design-construction integration is fundamental to ensure that a project will provide the service for which it was undertaken. A fundamental activity to reach this integration is the ability to write daily reports that record the design-construction interaction, which can prevent unnecessary risks and determine the success of the project.

Remember, the interaction starts with you. Engaging your day with the right attitude combined with the certainty that you know the work requirements will help you to earn three things: respect, authority, and the ability to inspire others in doing the right thing. In no way am I asking you to assume a "good guy attitude." That is, in fact, one of the greatest mistakes young inspectors make.

Lesson 2

Understanding the Problem
This Fieldbook Solves

Maintaining complete, comprehensive, detailed records of every process or deliverables items is not only a good idea, it is an unstated requirement that the inspector or supervisor is expected to achieve. It is essential to the efficient control of the work, to the achievement of all company and project objectives, and to the management of the potential risk and opportunities that could be encountered.

The Daily Field Report is an essential documentation mechanism at the job site in which the inspector, supervisor, or lead person has to indicate not only what did or did not happen on a given day or throughout a particular period but also if the defined procedure was followed and if the deliverables received met the requirements stated on the contract documents. A daily report is a tool that an inspector, supervisor, or lead person uses to document how the project is developing in compliance with contractual documents and serves as a basis to analyze production rates.

In the fieldbook, detailed information is recorded in such a manner that it may be consulted to confirm the particulars of the facts whenever any portion of the work is questioned or the performance of any party becomes an issue. The information also becomes the basis for support in the prosecution or defense of claims in both directions involving all the stakeholders

Companies take a great risk if they neglect the field reports because they don't recognize their value until there is a problem. Companies are taking a great risk if they fail to train their inspector, supervisor, or lead people in recognizing that one of the greatest issues with daily field reports is that the information in them is needed only when there is a problem—but when and if a problem arises, the information logged in the reports is of great importance in dealing with the problem.

The better the notes, the easier it is to deal with these problems. For example, during the installation of a pair of steel girders in one of my projects, the contractor's supervisor was having problems installing the girders. He suspected that the bearing seats on the cap in the middle bent were too low, and that was keeping the end of the girder from resting on the upper bent as it was supposed to. Their surveying team confirmed that the whole bent was about two feet low.

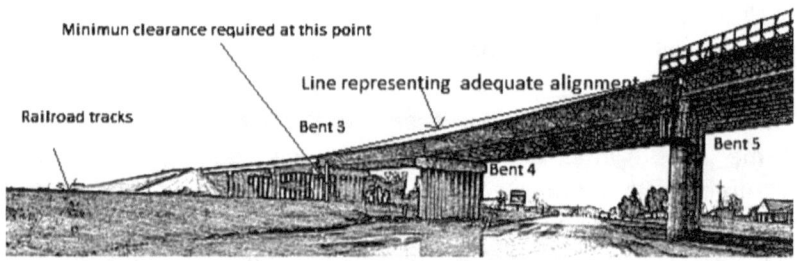

At that point I asked them to verify if the bent elevation was built according to plans and if the clearance on the railroad tracks would still be met as they had stated before. They afirmed that the bents numbered 3, 4, and 5 were built according to plans and that the clearance at the railroad tracks would be met despite this problem. They decided to proceed with the installation of the remaining girders and concrete beams.

When they installed the last girder three days later, I went to verify if the clearance at the railroad track was met. It was not. The cost for removing those beams and raising the structure was around $2 million.

Who was legally responsible for this additional cost? That's a question to be answered in another book. The main thing here is that I identified that the situation required detailed information to be written on the daily report to transfer the risk of continuing with the work to the contractor or at least to mitigate the risk of being legally responsible for the additional cost. My note was long, but here is part of it:

> I went back and met with the contractor's supervisor and recommended him to halt the girders installation until this matter was reviewed by the designer. He stated that upper management at his company instructed him to continue with the installation of the girders, since there was nothing that they could do. I told him to consider that any additional cost for having to remove those girders would be at their own expense, since a problem had been identified and needed to be resolved before they continued.

As I stated before, the contractor chose to continue with the installation. They accepted the risk of incurring on additional cost when they proceeded to perform an activity without obtaining proper documentation authorizing them to continue with the work despite the problem. A properly trained supervisor would write something like this on his report:

> I met with the lead inspector of the project to discuss the designer error found on the bearing seats elevations at bent 4. He has requested that the installation of the remaining girders be halted to allow the designer to review the problem and recommend an action to proceed. I have explained that halting the operation will cause an additional cost for lost time and remobilization of crews and equipment. Therefore we reserve our right to request compensation for all the additional costs and construction schedule delay. Later I contacted upper management and explained my conversation with the lead inspector. Upper management instructed me to continue with the installation.

A note like that would have been of real value to the supervisor and the supervisor's company. But, evidently, that decision was beyond the supervisor's authority. However, that note on his daily report would have helped his company to resolve a problem, because it is the contractor's responsibility to prove or justify all claims and requests in a timely manner. Besides, the supervisor might not have known if upper management had notified the engineer in writing about the intent to see compensation for the extra time and money. They also might have obtained proper authorization to proceed with the work when he was instructed to continue with the installation process. In any case, he was off the hook because he did his job by documenting the problem, avoiding unnecesary risk when he followed the lead inspector's directions and relayed the intent to request additional compensation once there was knowledge of the basis for the request.

Unfortunately, the creation of daily reports that lack real value is a common practice. For example, a less-experienced inspector that is not properly trained in identifying risk and writing daily reports would write, "Contractor continued placing steel girders on bent 3 and 4." This note would not help the inspection company to transfer or mitigate the risk of the contractor proceeding to perform unauthorized work. In addition, a supervisor that is not properly trained in identifying risk and writing daily reports would write nothing at all.

The inspector, supervisor, or lead person then needed to begin his or her duties with the profound realization that the reporting responsibility is a fundamental requirement of the position. Not doing it correctly is a major performance deficiency because of the critical nature of the information. This is a huge risk that could be avoided with the adequate use of daily field reports.

We need to accept the idea that there is a high probability that most of the information written on such reports will never be consulted, because there are no problems. But we must face the reality that when there is a problem, very detailed information will be needed. Since there is, of course, no way to determine in advance the specific information that will be needed in the case that a negative event occurs, it is crucial to maintain all the information in an acceptable degree of detail. Examples of negatives events that could require detailed reports include an embankment failing a field density test, failing to meet concrete strength requirements, an appearance of shrinkage cracking on decks, or an additional use of concrete due to inadequate grading on a bonding breaker. Training is important to ensure that workers know how to keep useful records

Lesson 3

Key Factors in Writing Effective Daily Reports

Remember that people are the most important assets in organizations and projects. Upper management must have the ability to develop and manage a team. Good teams do not develop by chance. We rely on the ability to build cooperative relationships among our people.

For that reason, it is important that senior management establishes clear guidelines for daily reports and emphasizes their importance to the supervisory team. In this way, the process of writing appropriate reports is not looked upon by the supervisors or those who prepare the reports as something the home office or senior management is not concerned about. It is also important that the project manager reviews the daily reports and the communication that he has with the team members that allows him to monitor and improve the content of the reports.

Many companies realize that one of the key factors for their success is the acquisition and retention of qualified people and people willing to be trained when mismatches between resposibility and competency are identified. These are important because, when time-consuming activities around the site begin to steal precious time from the inspector, supervisor, or lead person, it is tempting to simplify the information on the report.

If allowed, the reporting process will continue to degenerate. First, reports won't be prepared daily; then the supervisor will attempt to get caught up near the end of the week or at some other later period, compromising the accuracy of the information. The notes themselves will become short and marginally useful.

Qualified people that are willing to be trained to acquire the required competency are more likely to be aware of and understand the purpose of writing appropiate daily reports and submitting them on time. Becoming familiar with the requirements and significance of the contract, plans, specifications, and procedures helps supervisors and inspectors be firm yet friendly when arguing about the activity being performed.

Most importantly, this knowledge allows us to document when the other party is not performing within the contract agreement. In the construction industry, it is very common to find inspectors that were construction supervisors in a previous job. The purpose of writing daily reports from an inspector's perspective is slightly different than the purpose from a construction supervisor's perspective. It is important to understand their main similarities and differences. They are similar because both of them daily track the equipment on-site, the manpower on-site, the work performed, the measurements and calculations of the amount of work done during the day, and the evaluations of the activities executed. But they are different because, while the construction supervisor is coordinating and evaluating the necessary activities to build the project or items to be delivered, the inspector is evaluating for the correctness and completeness of procedures and the items delivered by the contractor so these can be approved and signed off by the project sponsor or customer.

Lesson 4

Guidelines for Inspectors in Writing Daily Reports

The inspector's primary responsibility is to see that work assigned to him is executed in accordance with the plans, specifications, and addendums, except for variations that are permited in writing by a superior. When such instructions are not received in writing, the inspector must include the change on his daily report, which his superior will approve. The inspector should write his observations throughout the work, noting all warnings and instruction given to the contractor. He must write his report with the idea in mind that after a structure has been accepted, it may be too late to hold the contractor responsible.

A boss of mine once told me, "If an inspector does not see that the work is done properly, it probably will not be done correctly." I agree but also disagree with such statement. I prefer to share with people that trust is always earned through goodwill, respect, tact, competence, a sense of fairness, and consistency based on the requirments and the intent of the contract. However, in the past I have seen many contractors try to diminish, equivocally, the inspector's authority. I have included the following paragraphs to remind you of the importance of the role and the authority and limitations that inspectors have on the highway construction.

Federal regulations apply to all federally funded projects. The degree to which TxDOT monitors these projects depends primarily on a combination of the funding source, the highway system, the statutory requirements, and the potential risk posed by noncompliance. The Texas Department of Transportation minimizes monitoring on projects off the state system that do not have federal or state funds. In this case, the local government assumes more responsibility for compliance with statutes. The monitoring of projects focuses on proper design applications of TxDOT material and construction quality standards.

In summary, TxDOT tailors the level of monitoring to the relative risk to TxDOT—keeping on mind its stewardship responsibilities. Through the LGPP transportation program, local goverments may contract consultants to provide construction management services. These consultants adopt the role of TxDOT during the construction of the project, but TxDOT oversees and signs off for final acceptance. Local governments include municipalties, counties, regional mobility authorities, local toll authorities, and even include private entities. The Texas Department of Transportation's specifications clearly establish rules and grant authority to inspectors. Consider the following:

- Inspectors are authorized representatives of engineers authorized to examine all work performed and materials furnished, including preparation, fabrication, and material manufacture.
- Inspectors inform the contractor of failures to meet contract requirements.
- Inspectors may reject work or materials and may suspend work until any issues can be referred to and decided by the engineer.
- Inspectors cannot alter, add, or waive contract provisions, issue instructions contrary to the contract, act as foremen for the contractor, or interfere with the management of the work.
- Work performed without suitable inspection, as determined by the engineer, may be ordered removed and replaced at the contractor's expense.

So, contractors need to clearly understand that an inspector is empowered by the State Department of Highways to enforce the specifications concerning the quality of the materials and the work performed by them.

Understanding the difference between the construction phase and the project life cycle helps us to realize why daily reports can start at any time prior to the agreed-on work start date of the construction phase and continue after the end of the construction phase. The construction phase is part of the project life cycle, which is a collection of generally sequential and sometimes overlapping project phases. Project phases are divsions within a project where extra control is needed to manage the completion of a major deliverable effectively. Daily reports are useful in all the project phases. For example, a transportation project can be mapped to the following life-cycle structure:

- Phase 1: preparing the business case and request for proposal and selecting the consulting firm that will provide the construction management and funding the project
- Phase 2: organizing and preparing plans, specs, bids, and the selection of general contractor
- Phase 3: construction begins and delivers the highway, streets, or facilities
- Phase 4: closing the project, culminating in archiving project documents

Reports should be input whether the inspectors are working in the field or in the office. If work was done in the office, the inspectors must state what work was done.

Reports should state whether lab consultants were present, performing material testing, and should include the particular tests performed. Do not include the test results information. Instead, note if the results was satisfactory or not, and enter the results of the test in the corresponding log book for further analysis. (See *Field Book for Quality Control in Earthwork Operations* at www.cs4highway.com.)

Pictures can be attached, but we should ask ourselves what their purpose is and identify if the work in the picture meets or does not meet specifications.

Do not leave unresolved issues open. If during inspection something is found to be out of specification, state the name of the supervisor that was informed and actions that were to be followed to close the issue.

Be objective on the information provided, report all facts, and record all relevant information. Keep frustrations, innuendos, and other remarks that do not belong in a professional communication out of it altogether. Take into consideration that government personnel may review documents and that all documents are subject to revision.

Draw only appropriate conclusions. Do not make speculative conjectures. Include only conclusions that are the result of a direct cause-and-effect relationship and that lead to or require some action (correction of work, backcharge, additional costs, etc.)

Being brief is acceptable, if what is written is complete and accurate. Outlined statements are fine, if they still include all the facts and complete descriptions.

Be precise. Note specific locations, limits of work, and whatever information is necessary to clearly identify the work and the processes described.

Do not lie. Never write that you inspect something when you didn't.

Identify all sources of information. Any information other than your own observations must have its source(s) identified. Refer to companies and name names.

Lesson 5

Guidelines for Contractors in Writing Daily Reports

The main reason a construction supervisor writes reports is to provide information to upper management about actual production rates, crew sizes, unexpected risks, better construction practices, and construction procedures that will help with estimating future jobs. Of course, the daily information allows project managers to determine variances when comparing the actual construction cost against the budgeted cost. Also, the fieldbook becomes an important source of information to determine the root causes of construction delays.

Construction supervisors need to identify and document factors that inhibit or reduce productivity. We all know that sometimes things happen out on the job that cause the tradespeople to lose time. It is therefore the responsibility of the construction supervisor to identify unproductive situations and to take corrective actions to prevent them in the future—by choice and not by coincidence. Experts who studied productivity on construction problems found that the most significant problems came from the following areas:

- Unavailability of construction materials and/or equipment
- Disrespectful treatement of workers
- Breakdown in communication
- Incompetent personnel or supervisors
- Unavailabilty of tools needed to do the work
- Project confusion and unsafe conditions
- Work that needed to be redone

Lesson 6

Design-Construction Narrative

This section describes the *focal points* that are considered when writing daily reports.

- *Weather statements* indicate the actual weather conditions in the project, including the minimum and maximum temperatures. Also indicate the estimated amount of water received during a precipitation event and any other weather conditions that halts the construction activities. If the activities are halted, be sure that the date is logged on the Lost Days section in the fieldbook.
- *Results of surveillance* include satisfactory work completed or deficiencies with action to be taken. Be observant and note what the critical activity is to ensure that you are attending to the most important matters. The critical activity of the day is not necessarily an item in the critical path, but could be.
- *Include tests required by the plans and/or specifications performed.* Do not indicate the result of the test. Instead, note whether the result was satisfactory or not, and enter the result of the test in the corresponding log book for further analysis. It is easier to analyze data when all the information for the analysis is in one place instead of flipping through and skipping pages to find it.
- *Important verbal instructions that were received or given.* State the method in which you gave or received the instruction. Cover any conflicts in plans, specifications, or instructions or any delay in the job, any decision-making discussion with the contractor, and any disagreements between parties, or other construction conflicts.
- *Give names of official visitors and a summary of discussions.* Official visitors are persons or organizations such as customers, sponsors, judges, commisioners, TxDOT

representatives, federal administrators, or auditors that are actively involved in the project or whose interest may be positively or negatively affected by the execution of it. It is critical for project success to identify them early in the project and analyse their level of power and interest to maximize positive influences and mitigate potential negative impacts.

- *Record results of safety inspections and unusual construction work conditions,* including safety violations and corrective action taken.
- *Measurements and calculations* show the amount of work the contractor has completed and must be recorded daily— unless the complexity or volume is not appropriate to do it in this way, such as with embankments and roadway excavations. These activities should be done in appropriate intervals established for the activity being done in periods of no more than fifteen days. Use the bid item numbers included on the contract to identify the activities being performed. For example, item 530-2010 refers to concrete driveways. If the contractor worked on this item, you should measure and calculate the amount of square yards completed. This information is compiled and used to determine the monthly estimate of work completed and the monthly payment to the contractor. These reports are audited frequently by TxDOT and federal government.
- *Include rework or field errors,* and identify if the work being performed is the result of poor workmanship or error.
- *For warranty work done,* identify if the work is the result of a contract warranty, and describe the possible causes of the failure.
- *For customer adjustments,* identify if the customer is requesting to perform work that is not included in the scope of work.
- *Estimate errors*: identify if the work done is included on the contract but not considered by the estimating team.

With today's increasing reliance on technology, daily reports commonly need to be submitted electronically. However, filling out electronic reports directly on-site is rarely practical or possible due to the nature of our work. This fieldbook is designed to be carried wherever you go. In this way you have it handy to write down all relevant information as it is observed. As you develop this habit, the daily completion of the electronically submitted report becomes almost automatic. At the end of the day, you have all the information you need to submit.

In conclusion, this *Highway Construction and Inspection Fieldbook* will help in your quest to

- provide a chronological account of the work actually performed;
- provide easily retrievable information;
- compile the amount of resources utilized during the construction of the project;
- record the sequence of events that led to a particular problem and document its solutions;
- document warnings and request for materials, information, or other assistance from upper management; and
- provide training and consistency in matters about writing daily reports.

Lesson 7

Calculating Project Durations
and Actual Working Days

Day of Mo.	Jan.	Feb.	Mar.	Apr.	May.	Jun.	Jul.	Aug.	Sep.	Oct.	Nov.	Dec.	Day of Mo.
1	1	32	60	91	121	152	182	213	244	274	305	335	1
2	2	33	61	92	122	153	183	214	245	275	306	336	2
3	3	34	62	93	123	154	184	215	246	276	307	337	3
4	4	35	63	94	124	155	185	216	247	277	308	338	4
5	5	36	64	95	125	156	186	217	248	278	309	339	5
6	6	37	65	96	126	157	187	218	249	279	310	340	6
7	7	38	66	97	127	158	188	219	250	280	311	341	7
8	8	39	67	98	128	159	189	220	251	281	312	342	8
9	9	40	68	99	129	160	190	221	252	282	313	343	9
10	10	41	69	100	130	161	191	222	253	283	314	344	10
11	11	42	70	101	131	162	192	223	254	284	315	345	11
12	12	43	71	102	132	163	193	224	255	285	316	346	12
13	13	44	72	103	133	164	194	225	256	286	317	347	13
14	14	45	73	104	134	165	195	226	257	287	318	348	14
15	15	46	74	105	135	166	196	227	258	288	319	349	15
16	16	47	75	106	136	167	197	228	259	289	320	350	16
17	17	48	76	107	137	168	198	229	260	290	321	351	17
18	18	49	77	108	138	169	199	230	261	291	322	352	18
19	19	50	78	109	139	170	200	231	262	292	323	353	19
20	20	51	79	110	140	171	201	232	263	293	324	354	20
21	21	52	80	111	141	172	202	233	264	294	325	355	21
22	22	53	81	112	142	173	203	234	265	295	326	356	22
23	23	54	82	113	143	174	204	235	266	296	327	357	23
24	24	55	83	114	144	175	205	236	267	297	328	358	24
25	25	56	84	115	145	176	206	237	268	298	329	359	25
26	26	57	85	116	146	177	207	238	269	299	330	360	26
27	27	58	86	117	147	178	208	239	270	300	331	361	27
28	28	59	87	118	148	179	209	240	271	301	332	362	28
29	29		88	119	149	180	210	241	272	302	333	363	29
30	30		89	120	150	181	211	242	273	303	334	364	30
31	31		90		151		212	243		304		365	31

In leap year, after February 28, add 1 to the tabulated number

Practical Example:

Project start date is	25-Jun	=	176
Project completion date is	12-Nov	=	316
Total Project duration is 316 -176 =	140	days	
Lost days recorded =	25	days	
Actual working days =	115	days